BANTAMS AND MINIATURE FOWL IN COLOUR

by

Michael Roberts

Illustrated by

Sara Roadnight

Design & Layout

Philip Drury
www.stunningmedia.co.uk

Photographed by

Grant Brereton
Richard Roadnight
& Eternal Photography

Introduction

This book has evolved from Bantams in Colour which was first published in 1984. We have updated that book and renamed it Bantams and Miniature Fowl in Colour; this is because the word bantam was misleading as most of the birds in the book are miniature fowl, that is to say they have large fowl counterparts, whereas a true bantam like the Sebright has no large fowl counterpart. The word bantam comes from Indonesia where there is a port in the north west of Java called Bantam from which sailing boats laden with spices set out for Europe.

We have kept most of the original photographs, which were taken by Richard Roadnight, but we are also grateful for the use of more up to date imagess which have been supplied by Eternal Photographs.

This book is sold on condition that none of the photographs can be used or reproduced in any way, shape or form.

Michael Roberts and Sara Roadnight,
Kennerleigh, 2010.

Contents

Ancona

Ancona (f)

Breed	Ancona
Type	Miniature
Classification	Light
Origin	Italy
Colour variations	Mottled
Egg colour	White
Specifications	Upright bearing, single or rose comb, white ear lobes, legs yellow but mottled with black. If single comb, in the male it should be erect and in the female it should flop over to one side. Excellent egg layer, but sometimes nervous.
Faults to avoid	Oversize.

Ancona (m)

Ancona (f)

Andalusian

Andalusian (f)

Breed	Andalusian
Type	Miniature
Classification	Light
Origin	Spain
Colour variations	Blue
Egg colour	White
Specifications	Upright bearing, single comb, white ear lobes. In the male the comb should be well serrated and erect and in the female it should flop over. Legs dark grey to black. Feathers blue with black lacing on body. Excellent egg layer, can be flighty.
Faults to avoid	Lack of lacing.

Andalusian (m)

Araucana

Rumpless Araucana (f)

Breed	Araucana
Type	Miniature
Classification	Light
Origin	Chile and Peru
Colour variations	Lavender, black, brown, any Old English Game colour
Egg colour	Blue
Specifications	This breed also has a rumpless variation. Legs are dark grey to black. Pea comb, very small in the female. There should to be a small crest, not upright, muffing. If rumpless there is no crest but long ear tufts. Good egg layer, hardy.
Faults to avoid	Oversize, lack of muffing.

LavenderAraucana (m)

Australorp

Blue Australorp (f)

Breed	Australorp
Type	Miniature
Classification	Heavy
Origin	England and Australia
Colour variations	Black, blue, white
Egg colour	Cream to light brown
Specifications	A stocky breed with single comb and black legs. Red ear lobes. Compact but upright tail. Good egg layer.
Faults to avoid	Long flowing tail.

Black Australorp (f)

Black Australorp (m)

Barbu d'Uccle

Breed	Barbu d'Uccle
Type	Bantam
Origin	Belgium
Colour variations	Cuckoo, black mottled, black, blue quail, lavender, lavender mottled, laced-blue, millefleur, porcelaine, white, quail
Egg colour	Cream
Specifications	Short, pert bearing, single comb, heavily feathered legs and feet. Also muffs and beard.
Faults to avoid	Incorrect comb and poorly feathered legs and feet.

Black Mottled Barbu d'Uccle (m)

Millefleur Barbu d'Uccle (f)

Millefleur Barbu d'Uccle (m)

Barbu d'Anvers

Black Barbu d'Anvers (m)

Breed	Barbu d'Anvers
Type	Bantam
Origin	Belgium
Colour variations	Cuckoo, black mottled, black, quail, blue quail, lavender, lavender mottled, laced-blue, millefleur, porcelaine, white.
Egg colour	Cream
Specifications	Upright, rose comb and clean legs. Wings held nearly vertically in the male. Bearded and muffed.
Faults to avoid	Single comb, large wattles and ear lobes, feathered legs.

Mottled Barbu d'Anvers (f)

Quail Barbu d'Anvers (m)

Blue Laced
Barbu d'Anvers (m)

Quail Barbu d'Anvers (f)

White Barbu d'Anvers (m)

Millefleur Barbu d'Anvers (f)

9

Barbu de Watermael

Cuckoo Barbu de Watermael (f)

Breed	Watermael
Type	Bantam
Classification	Soft feather
Origin	Belgium
Colour variations	Black, blue, blue quail, quail
Egg colour	Creamish white
Specifications	Rose comb not too pro-nounced, with a small crest and neat muffs. Must be a small dainty and active bird.
Faults to avoid	Oversized crest, single combe, eyes obscured coarseness, oversize, yellow legs.

Quail Barbu de Watermael (f)

Croad Langshan

Breed	Black Croad Langshan
Type	Miniature
Classification	Heavy
Origin	China
Colour variations	Black, white rarely
Egg colour	Light brown
Specifications	Single comb, lightly feathered legs and feet, greeny tinge to black plumage, white to pink legs. Proud carriage.
Faults to avoid	White ear lobes, no feathers or too large feathers on legs, yellow legs.

Black Croad Langshan (f)

Black Croad Langshan (m)

Brahma

Buff Columbian Brahma (f)

Breed	Brahma
Type	Miniature
Classification	Heavy
Origin	India
Colour variations	Light, dark, white, gold, birchen
Egg colour	Cream to light brown
Specifications	Upright, stocky and strong front, pea comb, feathered legs and feet, legs yellow in colour.
Faults to avoid	Too large a tail and legs too long. White ear lobes, clean legs.

Gold Brahma (f)

Light Brahma (f)

Birchen Brahma (f)

Gold Brahma (m)

Light Brahma (m)

Dark Brahma (m)

Dorking

Silver Grey Dorking (f)

Breed	Dorking
Type	Miniature
Classification	Heavy
Origin	England
Colour variations	Silver grey, red, white, cuckoo
Egg colour	Cream
Specifications	Single comb in red and silver grey, rose comb in cuckoo and white. Five toes. Boat-shaped body, regal carriage. White legs and feet. Wings horizontal with the ground and long saddle hackles.
Faults to avoid	Four-toed birds, incorrect comb, upright and short body.

Silver Grey Dorking (m)

Silver Grey Dorking (f)

Dutch

Gold Dutch (m)

Silver Dutch (m)

Silver Dutch (f)

Gold Dutch (f)

Breed	Dutch
Type	Bantam
Origin	Holland
Colour variations	Silver, gold, salmon, blue, partridge
Egg colour	Cream
Specifications	Single comb, upright, pert bearing, legs slate coloured, white ear lobes, wings held vertically, full tail, short back.
Faults to avoid	Red ear lobes, yellow legs, narrow build.

Faverolles

Black Faverolle (m)

Ermine Faverolle (f)

Salmon Faverolle (f)

Salmon Faverolle (m)

Breed	Faverolle
Type	Miniature
Classification	Heavy
Origin	France
Colour variations	Black, salmon, ermine, blue, buff, white
Egg colour	Cream
Specifications	Single comb, upright bearing, five toes, beard and side muffs, lightly feathered legs. Red ear lobes.
Faults to avoid	Lack of fifth toe, clean legs, no beard or muffing.

Salmon Faverolle (m)

Black Leghorns

White Leghorns

Frizzle

White Frizzle (m)

Breed	Frizzle
Type	Miniature
Classification	Light
Origin	China
Colour variations	Black, white, buff, blue, silver grey
Egg colour	Cream
Specifications	Each feather curling towards the head of the bird. Single comb, red ear lobes, legs yellow.
Faults to avoid	Lack of even curl in feathers, longtail.

Black Frizzle (f)

Buff Frizzle (f)

German Langshan

German Langshan (f)

Breed	German Langshan
Type	Miniature
Origin	Germany
Colour variations	Black, white, blue
Egg colour	Light brown to brown
Specifications	Small, single comb, with red face, wattles and ear-lobes, plumage beetle green black, snow white, or Andalusian-blue, an upright and graceful bird. Long un-feathered legs.
Faults to avoid	Large comb, white ear-lobes, short or feathery legs, imbalance

German Langshan (m)

German Langshan (f)

Hamburgh

Gold Spangled
Hamburgh (m)

Breed	Hamburgh
Type	Miniature
Classification	Light
Origin	Europe
Colour variations	Gold pencilled, gold spangled, silver pencilled, silver spangled
Egg colour	White
Specifications	Rose comb, white ear lobes, upright and alert. Fine, elegant bearing. Black or white bantam Hamburghs are not recognised as separate colours due to the similarity to Rosecombs.
Faults to avoid	Incorrect comb.

Silver Spangled Hamburgh (m)

Silver Spangled Hamburgh (f)

Gold Pencilled Hamburgh (m)

Gold Pencilled Hamburgh (f)

Gold Spangled Hamburgh (m)

Silver Pencilled Hamburgh (f)

Indian Game

Breed	Indian Game (Cornish)
Type	Miniature
Classification	Heavy
Origin	Cornwall, England
Colour variations	Indian, jubilee
Egg colour	Cream
Specifications	Indian colour is chestnut, laced with greeny black. Jubilee colour is chestnut, laced with white. Very stocky and broad with thick yellow legs set well apart. Meaty looking. Pea comb and red ear lobes.
Faults to avoid	Incorrect comb, leggy, narrow-chested

Jubilee Indian Game (m)

Indian Game (f)

Ko Shamo

Ko Shamo (m)

Ko Shamo (m)

Breed	Ko Shamo
Type	Miniature
Classification	Hard Feather
Origin	Japan
Colour variations	Black, black-red, blue, buff, duckwing, spangled, cuckoo, white, and mottled.
Egg colour	Cream and tinted
Specifications	Comb, walnut or pea. Earlobes red, thick beak deep and curved, dewlap large, glaring eyes, must be upright in carriage, strongly built but not coarse. Strong legs and feet.
Faults to avoid	Any fineness, single comb. Oversize and overweight.

Japanese

Breed	Japanese
Type	Bantam
Origin	Japan
Colour variations	Black, blue, mottled, black-tailed white, grey, buff, cuckoo, black-tailed buff
Egg colour	Cream
Specifications	Very short, clean legs, single comb, long upright tail, wings carried low, red ear lobes.
Faults to avoid	Long legs, low and short tail.

Grey Japanese (m)

Black-tailed white Japanese (f)

Black-tailed white Japanese (m)

White Japanese (m)

Mottled Japanese (f)

Grey Japanese (m)

Frizzle Japanese (f)

Leghorn

Breed	Leghorn
Type	Miniature
Classification	Light
Origin	Italy
Colour variations	White, blue, black, brown, cuckoo, duckwing, barred.
Egg colour	White
Specifications	Tall breed, large single comb in male, female's comb must flop over. White ear lobes, yellow legs. Rose comb sometimes seen. Good egg layer.
Faults to avoid	Incorrect leg colour, erect comb in female.

Blue Leghorn (f)

Black Leghorn (f)

Brown Leghorn (f)

Blue Leghorn (f)

White Leghorn (f)

White Leghorn (f)

Brown Leghorn (f)

Barred Leghorn (m)

Malay

White Malay (m)   Black Malay (f)

Breed	Malay
Type	Miniature
Classification	Heavy
Origin	Asia
Colour variations	Black, white, spangled, pile, duckwing
Egg colour	Cream
Specifications	Very tall, leggy, wiry bird, walnut comb. Yellow legs and read ear lobes.
Faults to avoid	Lack of height, stockiness, incorrect comb.

Marans

Marans (f)

Breed	Marans
Type	Miniature
Classification	Heavy
Origin	Holland
Colour variations	Cuckoo
Egg colour	Dark brown
Specifications	Single comb, red ear lobes, white legs, stocky build. The name Marans was given to the chickens brought by Dutch drainage engineers to the region of Marans in France in the sixteenth century.
Faults to avoid	Yellow legs, white ear lobes, pale eggs.

Marans (m)

Marans (f)

Minorca

Black Minorca (f)

Breed	Minorca
Type	Miniature
Classification	Light
Origin	Spain
Colour variations	Black, white, buff, blue
Egg colour	White
Specifications	Tall, upright bearing, large single comb flops over in the female, large white ear lobes, black legs. White legs only in white colour variation. Good egg layer. Eggs large.
Faults to avoid	Yellow legs, small ear lobes, erect comb in female.

Black Minorca (m)

Black Minorca (f)

Modern Game

Pile Modern Game (m)

Breed	Modern Game
Type	Miniature
Classification	Heavy
Origin	England
Colour variations	Pile, birchen, brown-red, black-red, silver duckwing, gold duckwing, black, blue, white
Egg colour	Cream
Specifications	Small single comb, but customary to dub the male combs. Very long legs, tall, well-rounded muscular bird.
Faults to avoid	Lack of height, incorrect comb. Check leg colour matches corresponding colour variation.

Silver Duckwing Modern Game (f)

Brown-red Modern Game (m)

Birchen Modern Game (f)

Nankin

Nankin (f)

Breed	Nankin
Type	Bantam
Origin	China
Colour variations	Ginger-buff
Egg colour	Cream
Specifications	Single comb or rose comb, red ear lobes, white legs, male should be orange marmalade colour, hen slightly paler.
Faults to avoid	Lack of black main tail feathers.

Nankin (m)

New Hampshire Red

New Hampshire Red (f)

Breed	New Hampshire Red
Type	Miniature
Classification	Heavy
Origin	U.S.A.
Colour variations	Chestnut
Egg colour	Light brown
Specifications	Single comb, red ear lobes, yellow legs. Black-edged wings, main tail feathers black. Upright bearing, solid and broad.
Faults to avoid	Incorrect leg colour or comb, lack of black feathers.

New Hampshire Red (m)

New Hampshire Red (f)

Pile Old English Game

Cuckoo Old English Game

Old English Game

Crele Old English Game (f)

Breed	Old English Game
Type	Miniature
Classification	Light
Origin	England
Colour variations	Pile, crele, silver duckwing, wheaten, spangle, partridge, black-red, blue dun, white, black, furnace, cuckoo, ginger-red, henny, yellow duckwing
Egg colour	Cream
Specifications	Small single comb, but customary to dub male combs. Well-rounded, neat muscular bird, heart-shaped when viewed from above.
Faults to avoid	Oversize. Check leg colour matches corresponding colour variation.

Silver Duckwing Old English Game (m)

Silver Duckwing Old English Game (f)

Pile Old English
Game (m)

Spangle Old
English Game (m)

Wheaten Old
English Game (f)

Partridge Old
English Game (f)

Black-red Old
English Game (m)

Spangled (f)

Furnace (m)

Polecat (f)

Spangled (f)

Spangled (m)

Blue Furnace (m)

Orloff

White Orloff (m)

Breed	Orloff
Type	Miniature
Classification	Heavy
Origin	Russia
Colour variations	White, black, mahogany, spangled
Egg colour	Cream
Specifications	Pea comb, yellow legs, upright bearing, thick neck, solid looking, ear lobes red, small muff and beard.
Faults to avoid	Incorrect leg or comb, lack of muffs and beard.

White Orloff (f)

Black and Chocolate Orpingtons

Blue Orpingtons

Orpington

Buff Orpington (f)

Buff Orpington (m)

Black Orpington (m)

Black Orpington (f)

Breed	Orpington
Type	Miniature
Classification	Heavy
Origin	Kent, England
Egg colour	Light brown
Specifications	Compact shape, small single comb, red ear lobes, white legs, feathers soft and fluffy. Black variety may have a rose comb.
Faults to avoid	Oversize single comb, legginess.

Chocolate Orpington (m)

Chocolate Orpington (f)

Blue Orpington (f)

White Orpington (m)

Pekin

Breed	Pekin
Type	Miniature (of Cochin)
Classification	Heavy
Origin	China
Colour variations	Black, white, buff, cuckoo, mottled, blue, lavender, partridge, barred, columbian
Egg colour	Cream
Specifications	Single comb, red ear lobes. Should be circular in shape when viewed from above, whole out-line rounded, heavily feathered legs and feet.
Faults to avoid	Oversize, lack of feathers on feet.

Cuckoo Pekin (f)

Buff Pekin (m)

Lavender Pekin (f)

Columbian Pekin (f)

Columbian Pekin (m)

Blue Pekin (f)

Black Pekin (f)

Black Pekin (m)

Mottled Pekin (f)

Lavender Pekin (f)

Cuckoo Pekin (f)

Buff Pekin (m)

White Pekin (f)

Lavender Pekin (f)

Mottled Pekin (f)

Buff Pekin (m)

White Pekin (f)

Phoenix

Gold Duckwing Phoenix (m)

Breed	Phoenix
Type	Miniature
Classification	Light
Origin	Japan
Colour variations	Silver duckwing, gold duckwing
Egg colour	Cream
Specifications	Single comb, tail must be at least twice length of body in male, female has longish tail held horizontally. Legs yellow.
Faults to avoid	Pea comb. Beware confusion with Yokohama, a rare breed. In Britain the Phoenix is sometimes mistakenly called Yokohama.

Gold Duckwing Phoenix (f)

Silver Phoenix (m)

Silver Sebrights

Gold Sebrights

Plymouth Rock

Partridge Plymouth Rock (f)

Breed	Plymouth Rock
Type	Miniature
Classification	Heavy
Origin	U.S.A.
Colour variations	Barred, partridge, buff, white, black, columbian.
Egg colour	Cream
Specifications	Stocky, well-built bird, single comb, red ear lobes, yellow legs.
Faults to avoid	Incorrect leg colour or comb.

Barred Plymouth Rock (m)

Buff Plymouth Rock (m)

Barred Plymouth Rock (m)

Buff Plymouth Rock (m)

White Plymouth Rock (f)

Partridge Plymouth Rock (f)

Poland

Chamois Poland (f)

Breed	Poland
Type	Miniature
Classification	Light
Origin	Europe
Colour variations	White-crested black, black, white, gold, silver, white-crested blue, chamois
Egg colour	White
Specifications	Upright with profuse crest, compact in female, more open and spiky in male, uniform in colour. Small horn comb, white ear lobes.
Faults to avoid	Lack of crest, uneven shape and colour of crest. Check leg colour matches corresponding colour variation.

Silver Poland (m)

Silver Poland (f)

White-crested
Black Poland (f)

Black Poland (f)

Gold Poland (m)

White-crested Blue Poland (f)

Rhode Island Red

Rhode Island Red (f)

Breed	Rhode Island Red
Type	Miniature
Classification	Heavy
Origin	U.S.A.
Colour variations	Dark mahogany with beetle-green sheen
Egg colour	Light brown
Specifications	Single or rose comb, yellow legs, red ear lobes, solid carriage. Black-edged wings, main tail feathers black. Good egg layer
Faults to avoid	Incorrect leg colour, white ear lobes, light coloured plumage.

Rhode Island Red (m)

Rhode Island Red (f)

Rosecomb

Black Rosecomb (m)

Breed	Rosecomb
Type	Bantam
Origin	Asia
Colour variations	Black, white, blue
Egg colour	Cream
Specifications	Rose comb, largish white ear lobes. Legs black with black colour, white with white, slate with blue. Male is a graceful bird with long sickle tail.
Faults to avoid	Incorrect comb, lack of white ear lobes.

White Rosecomb (f)

Black Rosecomb (f)

Blue Rosecomb (m)

Rumpless Game

Pile Rumpless Game (m)

Wheaten Rumpless Game (f)

Breed	Rumpless Game
Type	Miniature
Classification	Light
Origin	England
Colour variations	As Old English Game
Egg colour	Cream
Specifications	Single comb, distinguishing feature of no parson's nose and no tail, otherwise similar to Old English Game, but not dubbed.
Faults to avoid	Oversize

Scots Dumpy

Breed	Scots Dumpy
Type	Miniature
Classification	Light
Origin	Scotland
Colour variations	Black, cuckoo
Egg colour	Cream
Specifications	Very short legs, long back with a waddling walk. Single comb, red ear lobes. Tail should be full and flowing.
Faults to avoid	Long-leggedness, five toes.

Scots Dumpy (f)

Scots Dumpy (m)

Scots Grey

Breed	Scots Grey
Type	Miniature
Classification	Light
Origin	Scotland
Colour variations	Cuckoo
Egg colour	Cream
Specifications	Single comb, white mottled legs, fine, compact, smart bird, with well-defined markings. Red ear lobes.
Faults to avoid	Heaviness, incorrect comb.

Scots Grey (f)

Scots Grey (m)

Sebright

Gold Sebright (f)

Breed	Sebright
Type	Bantam
Origin	England
Colour variations	Gold, silver
Egg colour	Cream
Specifications	Pert with tail carried high and prominent breast. Legs slate, rose comb, mulberry colour preferred. Correct lacing most important. Citron colour has been obtained by crossing gold and silver.
Faults to avoid	Single comb, fuzzy lacing.

Silver Sebright (m)

Silkies

White Silkie (m)

Breed	Silkie
Type	Miniature
Origin	Asia
Colour variations	White, black, blue, gold, partridge.
Egg colour	Cream.
Specifications	Mulberry comb with small prominences, small fluffy crest, and downy/fluffy body, black skin and legs with feathers. Five toes.
Faults to avoid	Oversize, plumage not Silky, wrong leg colour, four toes, and redness in face.

Partridge Silkie (f)

Black Silkie (f)

White Silkie (f)

Transylvanian Naked Neck

White Transylvanian Naked Neck (m)

Breed	Transylvanian Naked Neck (Turken)
Type	Miniature
Classification	Heavy
Origin	Hungary
Colour variations	White, black, blue, cuckoo, buff, red
Egg colour	Cream
Specifications	Single comb, featherless neck to crop, small cap of feathers surrounding comb. Supposed to have half the number of feathers as chickens of comparable size. Naked areas bright red.
Faults to avoid	Any feathers on neck.

White Transylvanian Naked Neck (f)

Sussex

Light Sussex (m)

Breed	Sussex
Type	Miniature
Classification	Heavy
Origin	Sussex, England
Colour variations	Light, speckled, buff, white, silver, brown, red
Egg colour	Cream to light brown
Specifications	Single comb, red ear lobes, white legs. Solid carriage. Good egg layer.
Faults to avoid	Legginess, yellow legs, too lightweight, lack of black hackle.

Buff Sussex (f)

Silver Sussex (m)

Light Sussex (m)

Light Sussex (f)

Silver Sussex (m)

Silver Sussex (f)

Speckled Sussex (f)

Buff Sussex (f)

Welsummer

Welsummer (f)

Breed	Welsummer
Type	Miniature
Classification	Light
Origin	Holland
Colour variations	Golden brown
Egg colour	Dark brown
Specifications	Single comb, red ear lobes, yellow legs. Note that the cock and hen have different makings. Compact bird. Good egg layer.
Faults to avoid	White legs, too much red in feathers.

Welsummer (m)

Welsummer (f)

Wyandotte

Barred Wyandotte (m)

Breed	Wyandotte
Type	Miniature
Classifications	Heavy
Origin	U.S.A.
Colour variations	Black, blue, blue-laced, buff, buff-laced, columbian, cuckoo, gold-laced, mottled, silver-laced, partridge, red, silver-pencilled, white
Egg colour	Cream
Specifications	A sturdy breed. Rose comb, red ear lobes, yellow legs. Good egg layer.
Faults to avoid	Oversized rose comb, incorrect leg colour, oversize.

Partridge Wyandottes

Barred Wyandotte (f)

Chocolate
Wyandotte (f)

White Wyandotte (f)

Blue Laced
Wyandotte (f)

Silver Laced
Wyandotte (f)

White
Wyandotte (m)

Partridge
Wyandotte (m)

Partridge
Wyandotte (f)

Blue Partridge (m)

Blue Partridge (f)

Columbian
Wyandotte (m)

Silver Pencilled
Wyandotte (f)

The different types of comb

LARGE SINGLE COMB
e.g. Ancona

WALNUT COMB
e.g. Malay

TRIPLE OR PEA COMB
e.g. Brahma

FOLDED SINGLE COMB
e.g. Leghorn (female)

ROSE COMB
e.g. Sebright

MULBERRY COMB
e.g. Silkie

HORN COMB

SMALL SINGLE COMB
e.g. Old English Game

Feather variations and markings as conforming to Poultry Club Standards

Neck hackle
of Black-Red
Old English Gamecock

Neck hackle
of Partridge
Wyandotte cock

Body feather
of Partridge
Wyandotte hen

Neck hackle
of Gold-Laced
Wyandotte cock

Body feather
of Gold-Laced
Wyandotte hen

Neck hackle of
Light Sussex cock

Neck hackle of
Speckled Sussex cock

Body feather of
Speckled Sussex hen

Neck hackle of
Blue Andalusian cock

Body feather of
Blue Andalusian hen

Body feather of Blue
Belgian Barbu d'Anvers

Body feather of
Rhode Island Red hen

Neck hackle
of Ancona cock

Body feather
of Ancona hen

Neck hackle
Plymouth Rock cock

Body feather
Plymouth Rock hen

Neck hackle
of Barred
Plymouth Rock cock

Body feather
of Barred
Plymouth Rock hen

Neck hackle
of Marans cock

Body feather
of Maran shen

Body feather of
Indian Game hen

Body feather
of Silver-Laced
Wyandotte hen

Body feather
of Australorp hen

Body feather
of White Silkie hen

Body feather
of Silver-Grey
Dorking hen

Body feather
of Brown
Leghorn hen

Body feather
of Gold Duckwing
Old Dutch

Egg colour variation

Dark Brown

Light Brown

Blue/Green

Cream

White

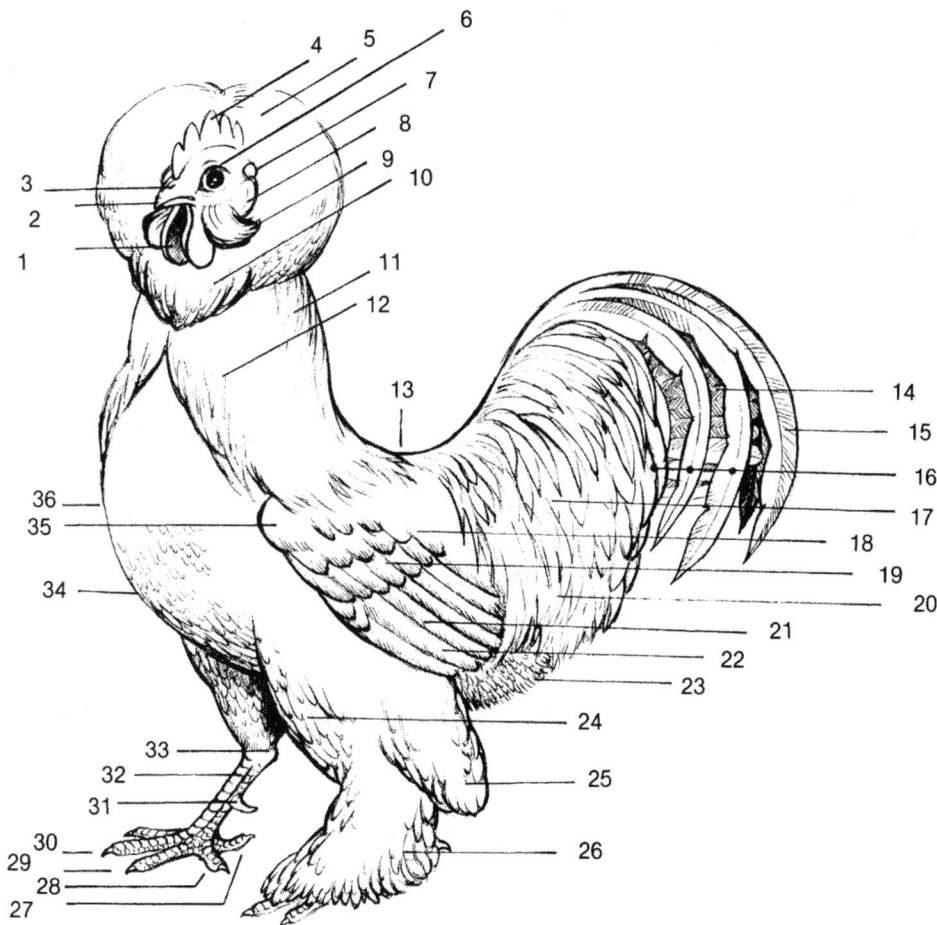

Labelled diagram of a chicken showing characteristics of most breeds

1 Wattles	10 Beard	19 Wing bar	28 Fourth toe
2 Beak	11 Neck	20 Saddle hackle	29 Third toe
3 Nostril	12 Neck hackle	21 Wing bay	30 Middle toe
4 Comb	13 Back	22 Primaries	31 Spur
5 Crest	14 Main tail	23 Abdomen	32 Shank
6 Eye	15 Sickle feather	24 Thigh	33 Hock joint
7 Ear	16 Side hangers	25 Vulture hock	34 Keel
8 Ear lobe	17 Tail coverts	26 Footings	35 Shoulder
9 Muffing	18 Wing bow	27 Fifth toe	36 Breast

Notes